SURVEYING MATHEMATICS MADE SIMPLE

An original book by

Jim Crume P.L.S., M.S., CFedS

Co-Authors
Cindy Crume
Bridget Crume
Troy Ray R.L.S.
Mark Sandwick L.S.I.T.

PRINTED EDITION

*****

PUBLISHED BY:

Jim Crume P.L.S., M.S., CFedS

Spiral Curves

Book 6 of this Math-Series

Copyright 2013 © by Jim Crume P.L.S., M.S., CFedS

All Rights Reserved

First publication: November, 2013

Printed by CreateSpace

Available on Kindle and other devices

TERMS AND CONDITIONS

The content of the pages of this book is for your general information and use only. It is subject to change without notice.

Neither we nor any third parties provide any warranty or guarantee as to the accuracy, timeliness, performance, completeness or suitability of the information and materials found or offered in this book for any particular purpose. You acknowledge that such information and materials may contain inaccuracies or errors and we expressly exclude liability for any such inaccuracies or errors to the fullest extent permitted by law.

Your use of any information or materials in this book is entirely at your own risk, for which we shall not be liable. It shall be your own responsibility to ensure that any products, services or information available in this book meet your specific requirements.

This book may not be further reproduced or circulated in any form, including email. Any reproduction or editing by any means mechanical or electronic without the explicit written permission of Jim Crume is expressly prohibited.

Table of Contents

INTRODUCTION..4

FULL TRANSITION SPIRAL CURVE - EQUAL TANGENTS...7

FULL TRANSITION SPIRAL CURVE - UNEQUAL TANGENTS..17

TEN CHORD SPIRAL - RAILROAD SPIRAL CURVE..29

POINTS ON A SPIRAL CURVE..............................32

SOLUTIONS TO EXAMPLES................................39

CONCLUSION...54

ABOUT THE AUTHOR......................................55

INTRODUCTION

Straight forward Step-by-Step instructions.

This book is just one part in a series of digital and printed editions on Surveying Mathematics Made Simple. The subject matter in this book will utilize the methods and formulas that are covered in the books that precede it. If you have not read the preceding books, you are encouraged to review a copy before proceeding forward with this book.

For a list of books in this series, please visit:

http://www.cc4w.net/ebooks.html

Prerequisites for this book:

A basic knowledge of geometry, algebra and trigonometry is required for the explanations shown in this book.

Book 1 - Bearings and Azimuths - How to add bearings and angles, subtract between bearings, convert from degrees-minutes-seconds to decimal degrees, convert from decimal degrees to degrees-minutes-seconds, convert from bearings to azimuths and convert from azimuths to bearings.

Book 2 - Create Rectangular Coordinates - How to calculate the northing and easting of an end point given the coordinates of the beginning point utilizing a bearing and distance of a line.

Book 3 - *Inverse Between Rectangular Coordinates* - How to determine the bearing and distance of a line given the coordinates for the beginning and ending point.

Book 4 - *Circular Curves* - How to calculate a circular curve, reverse curve, compound curve, Tangent In, Tangent Out and Local Tangent Bearing given only two parameters.

Definitions:

Spiral Curve: (a.k.a. Euler Spiral, Easement Curve, Transition Curve) A curve which the degree of curvature at any point is proportional directly to the distance the point is from the point of zero curvature. The Euler (Pronounced 'oil-er') Spiral curve is a clothoid with the cubic parabola having definite mathematical equations. This spiral curve is the accepted standard and has been utilized for highway and railroad curves since its inception by Leonhard Euler.

Spiral curves were originally designed for the Railroads to smooth the transition from a tangent line into simple curves. They helped to minimize the wear and tear on the tracks. Spiral curves were implemented at a later date on highways to provide a smooth transition from the tangent line into simple curves. The highway engineers later determined that most drivers will naturally make that transition with the vehicle; therefore, spiral curves are only used on highways in special cases.

Because they were used in the past and in special cases today, you still need to know how to calculate them.

FULL TRANSITION SPIRAL CURVE - EQUAL TANGENTS

Figure 1 & 2 shows the various components for a spiral curve. It is important that you become familiar with these components. They will be referenced throughout this book.

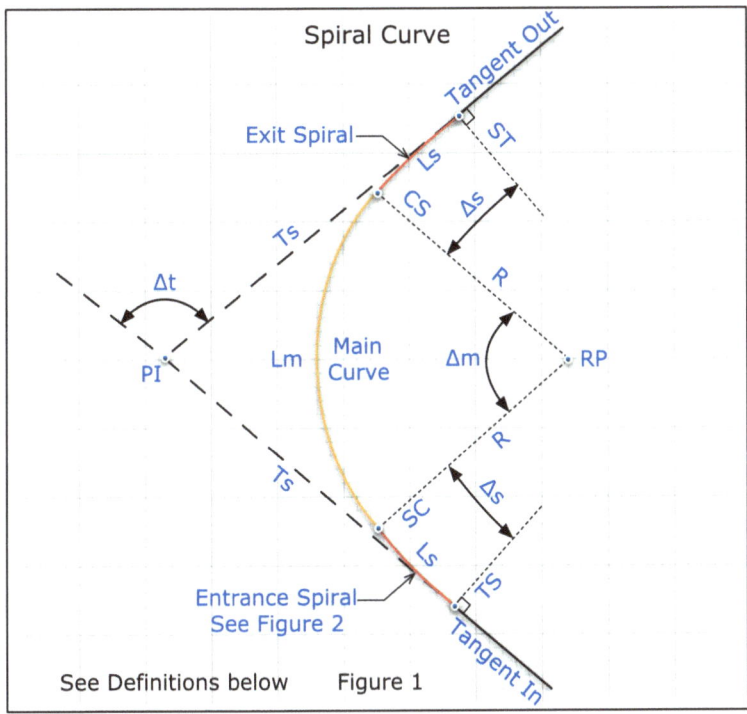

Figure 1

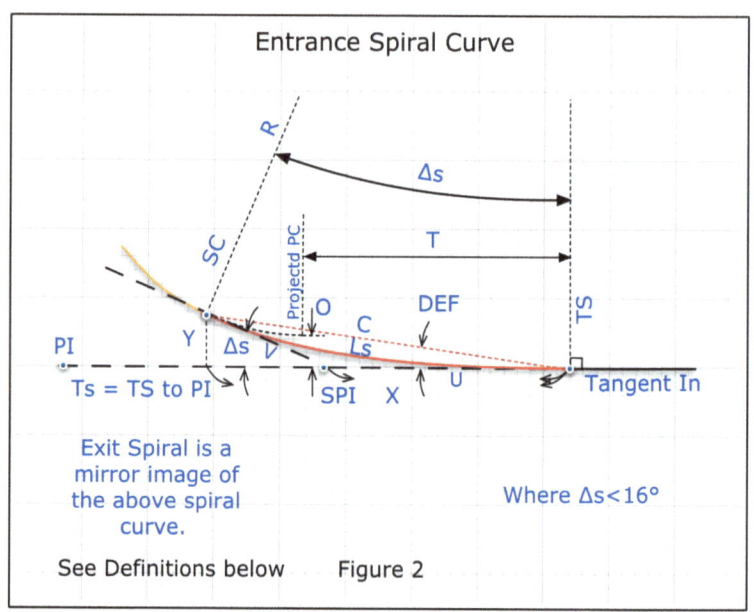

Figure 2

Definitions:

Δt = Total Delta Deflection
Δm = Delta of Main Curve
Δs = Spiral Delta
D = Degree of Curvature of Main Curve
Lm = Length of Main Curve
Ls = Length of Spiral
R = Radius of Main Curve
a = Rate of change per 100'
O = Radial Offset
T = Projected Curve P.C.
Ts = Tangent Length
C = Spiral Chord

DEF = Deflection angle at TS(Sta)
U = Distance from TS to SPI
V = Distance from SPI to SC(Sta)
X = Distance along X axis
Y = Distance along Y axis
TS(Sta) = Tangent to Spiral
SC(Sta) = Spiral to Curve
CS(Sta) = Curve to Spiral
ST(Sta) = Spiral to Tangent
PI(Sta) = Point of Intersection
RP = Radius Point of Main Curve
SPI = Spiral Point of Intersection

Formulas:

$R = 5729.57795 / D$

$a = (D * 100) / Ls$

$O = 0.0727 * a * ((Ls / 100)^3)$

$T = (Ls / 2) - (0.000127 * a^2 * (Ls / 100)^5)$

$Ts = (Tan(\Delta t / 2) * (R + O)) + T$

$C = Ls - (0.00034 * a^2 * (Ls / 100)^5)$

$DEF = (a * Ls^2) / 60000$

$\Delta s = 0.005 * D * Ls$

$U = C * Sin(\Delta s * 2 / 3) / Sin(\Delta s)$

$V = C * Sin(\Delta s * 1 / 3) / Sin(\Delta s)$

$\Delta m = \Delta t - \Delta s - \Delta s$

$L_m = (\Delta_m * R * \pi) / 180$

$X = C * \cos(DEF)$

$Y = C * \sin(DEF)$

$SC(Sta) = TS(Sta) + L_s$

$CS(Sta) = SC(Sta) + L_m$

$ST(Sta) = CS(Sta) + L_s$

$PI(Sta) = TS(Sta) + T_s$

Note: Rounding error is dependent upon the number of decimal places that are utilized. It is recommended that at least 5 decimal places be used for all calculations then round the final answer as needed.

All angles must be converted to Decimal Degrees prior to performing trigonometric operations. See Book 1 - "Bearings and Azimuths" for methods on converting Degrees-Minutes-Seconds to Decimal Degrees and vice versa. Also see Book 1 for adding and subtracting bearings and angles.

Example 1:

Refer to Figures 1 and 2 for spiral curve nomenclature.

Given:

$\Delta t = 36°29'16"$

$D = 2°00'00"$

$L_s = 200.00'$

$TS(Sta) = 2180+84.70$

Solve for the following elements:
—**SPIRAL CURVE**—

$R = 5729.57795 / D$
$R = 5729.57795 / 2°00'00"$
$R = \mathbf{2864.78898}$
$a = (D * 100) / Ls$
$a = (2°00'00" * 100) / 200.00$
$a = \mathbf{1.00}$
$O = 0.0727 * a * ((Ls / 100)^3)$
$O = 0.0727 * 1.00 * ((200.00 / 100)^3)$
$O = \mathbf{0.58160}$
$T = (Ls / 2) - (0.000127 * a^2 * (Ls / 100)^5)$
$T = (200.00 / 2) - (0.000127 * 1.00^2 * (200.00 / 100)^5)$
$T = \mathbf{99.99594}$
$Ts = (Tan(\Delta t / 2) * (R + O)) + T$
$Ts = (Tan(36°29'16" / 2) * (2864.78898 + 0.58160)) + 99.99594$
$Ts = \mathbf{1044.51462}$
$C = Ls - (0.00034 * a^2 * (Ls / 100)^5)$
$C = 200.00 - (0.00034 * 1.00^2 * (200.00 / 100)^5)$
$C = \mathbf{199.98912}$
$DEF = (a * Ls^2) / 60000$
$DEF = (1.00 * 200.00^2) / 60000$
$DEF = \mathbf{0.666667° \text{ or } 0°40'00"}$
$\Delta s = 0.005 * D * Ls$
$\Delta s = 0.005 * 2°00'00" * 200.00$

Δs = **2.00000° or 2°00'00"**

U = C * Sin(Δs * 2 / 3) / Sin(Δs)

U = 199.98912 * Sin(2°00'00" 2 / 3) / Sin(2°00'00")

U = **133.34112**

V = C * Sin(Δs * 1 / 3) / Sin(Δs)

V = 199.98912 * Sin(2°00'00" * 1 / 3) / Sin(2°00'00")

V = **66.67508**

—MAIN CURVE—

Δm = Δt - Δs - Δs

Δm = 36°29'16" - 2°00'00" - 2°00'00"

Δm = **32°29'16"**

Lm = (Δm * R * π) / 180

Lm = (32°29'16" * 2864.78898 * π) / 180

Lm = **1624.38889**

Note: See Book 4 - "Circular Curves" for formulas to solve other elements of the main curve.

X = C * Cos(DEF)

X = 199.98912 * Cos(0°40'00")

X = **199.97558**

Y = C * Sin(DEF)

Y = 199.98912 * Sin(0°40'00")

Y = **2.32693**

—STATIONING—

SC(Sta) = TS(Sta) + Ls
SC(Sta) = 2180+84.70 + 200.00
SC(Sta) = **2182+84.70**
CS(Sta) = SC(Sta) + Lm
CS(Sta) = 2182+84.70 + 1624.39
CS(Sta) = **2199+09.09**
ST(Sta) = CS(Sta) + Ls
ST(Sta) = 2199+09.09 + 200.00
ST(Sta) = **2201+09.09**
PI(Sta) = TS(Sta) + Ts
PI(Sta) = 2180+84.70 + 1044.51
PI(Sta) = **2191+29.21**

Note: The PI(Sta) is always calculated from the TS(Sta) and never from the ST(Sta).

Practical Example 1

At a minimum three (3) elements are required to calculate a full transitional spiral curve. [1] The Total Delta Deflection (Δt), [2] Degree of Curvature (D) for the main curve and [3] Spiral length (Ls). The Tangent to Spiral (TS) stationing is optional. This information can be derived from existing right of way plans, design plans and/or engineering design tables.

In order to calculate a full transitional spiral curve between two tangent lines, you first must determine the "Tangent In" and "Tangent Out" either by a survey of existing monumentation for existing centerlines or by a design alignment for a new centerline. It is beyond the scope of this book to describe the methods required to process existing monumentation to perform a centerline analysis for determination of the "Tangent In" and "Tangent Out".

For the purposes of this example, the "Tangent In", "Tangent Out", D, Ls and TS(Sta) will be provided.

Given:

Tangent In: S70°12'24"W

Tangent Out: N64°17'36"W

D = 2°30'00"

Ls = 300.00'

TS(Sta) = 10+00.00

Solve for the following elements:

$\Delta t = ??°??'??"$

$R = ????.?????$

$a = ?.??$

$Ts = ????.?????$

$C = ????.?????$

$DEF = ??°??'??"$

$\Delta s = ??°??'??"$

$\Delta m = ??°??'??"$

$Lm = ????.?????$

$SC(Sta) = ???+??.??$

$CS(Sta) = ???+??.??$

$ST(Sta) = ???+??.??$

$PI(Sta) = ???+??.??$

After the above items have been solved for, these values can be used to calculate coordinates for the key elements of the curve such as the PI, TS(Sta), SC(Sta), RP, CS(Sta), ST(Sta) and/or drawn in a CADD environment.

The solution can be found at the end of the book.

NOTES

FULL TRANSITION SPIRAL CURVE - UNEQUAL TANGENTS

Figure 3 & 4 shows the various components for a spiral curve with unequal tangents and spirals. It is important that you become familiar with these components. They will be referenced throughout this book.

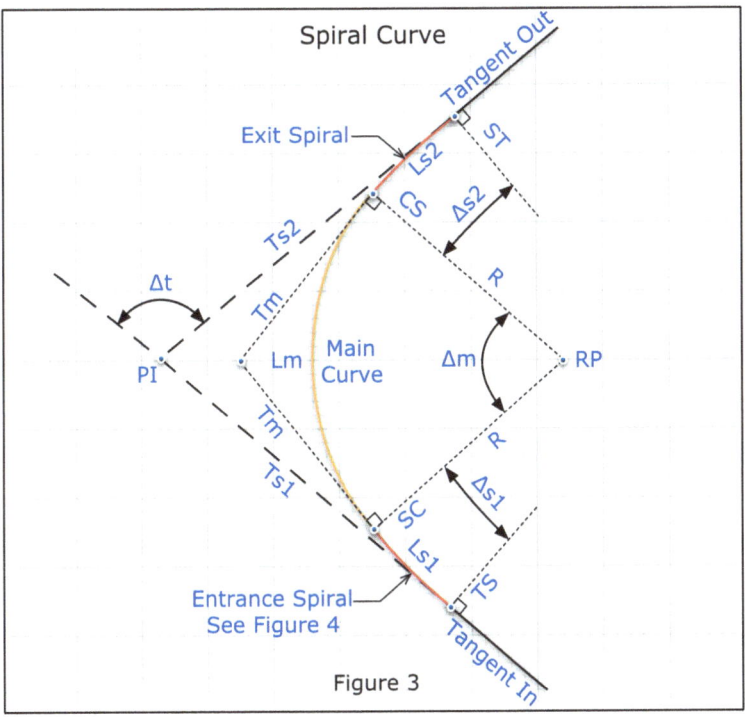

Figure 3

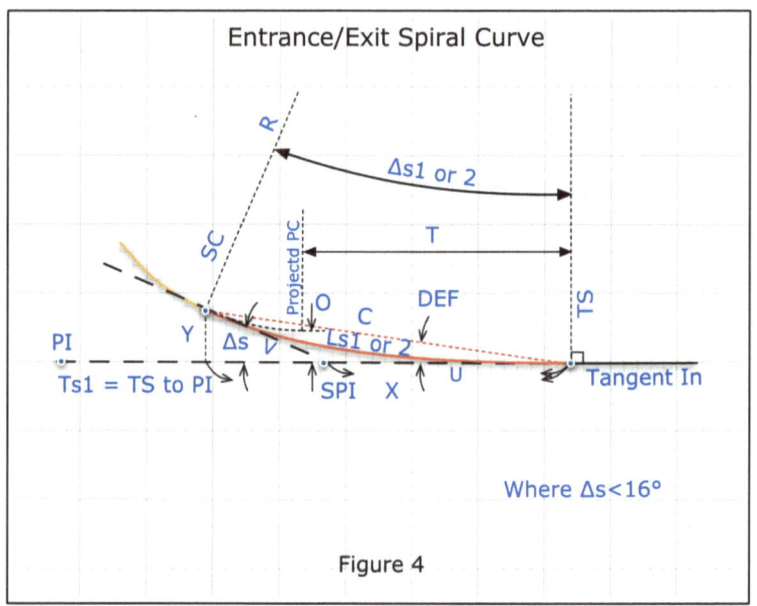

Figure 4

Occasionally, although rare, a centerline design may require an entrance spiral with one length and an exit spiral with a different length separated by a simple curve which results in two different tangent lengths for the overall spiral curve.

Refer to the definitions and formulas defined in the section "Full Transition Spiral Curve - Equal Tangents" that are in common with the following examples:

Additional Definitions:

Tm = Tangent distance for Main Curve

??1 = Variable related to Entrance Spiral

??2 = Variable related to Exit Spiral

Xa, Xb, Xc, Ya & Yb = Temporary variables

Additional Formulas:

$\Delta m = \Delta t - \Delta s_1 - \Delta s_2$
$X_a = \cos(\Delta s_1) * (V_1 + T_m)$
$Y_a = \sin(\Delta s_1) * (V_1 + T_m)$
$X_b = \cos(\Delta m + \Delta s_1) * (V_2 + T_m)$
$Y_b = \sin(\Delta m + \Delta s_1) * (V_2 + T_m)$
$X_c = (Y_a + Y_b) / \tan(\Delta t)$
$T_{s1} = X_a + X_b - X_c + U_1$
$T_{s2} = \sqrt{(X_c^2 + (Y_a + Y_b)^2)} + U_2$

Note: Rounding error is dependent upon the number of decimal places that are utilized. It is recommended that at least 5 decimal places be used for all calculations then round the final answer as needed.

All angles must be converted to Decimal Degrees prior to performing trigonometric operations. See Book 1 - "Bearings and Azimuths" for methods on converting Degrees-Minutes-Seconds to Decimal Degrees and vice versa. Also see Book 1 for adding and subtracting bearings and angles.

Example 2:

Refer to figures 1, 2, 3 and 4 for spiral curve nomenclature.

Given:

$\Delta t = 36°29'16"$
$D = 2°00'00"$
$L_{s1} = 200.00'$
$L_{s2} = 300.00'$

TS(Sta) = 2180+84.70

Solve for the following elements:

Entrance Spiral (Ls1)

a = (D * 100) / Ls1

a = (2°00'00" * 100) / 200.00'

a = **1.00**

O = 0.0727 * a * ((Ls1 / 100)³)

O = 0.0727 * 1.00 * ((200.00 / 100)³)

O = **0.58160**

T = (Ls1 / 2) - (0.000127 * a² * (Ls1 / 100)⁵)

T = (200.00 / 2) - (0.000127 * 1.00² * (200.00 / 100)⁵)

T = **99.99594**

C = Ls1 - (0.00034 * a² * (Ls1 / 100)⁵)

C = 200 - (0.00034 * 1.00² * (200.00 / 100)⁵)

C = **199.98912**

DEF = (a * Ls1²) / 60000

DEF = (1.00 * 200.00²) / 60000

DEF = **0.66667° or 0°40'00"**

Δs1 = 0.005 * D * Ls1

Δs1 = 0.005 * 2°00'00" * 200.00

Δs1 = **2.00000° or 2°00'00"**

U1 = C * Sin(Δs1 * 2 / 3) / Sin(Δs1)

U1 = 199.98912 * Sin(2°00'00" * 2 / 3) / Sin(2°00'00")

$U_1 = \mathbf{133.34112}$

$V_1 = C * \sin(\Delta s1 * 1 / 3) / \sin(\Delta s1)$

$V_1 = 199.98912 * \sin(2°00'00" * 1 / 3) / \sin(2°00'00")$

$V_1 = \mathbf{66.67508}$

$X = C * \cos(DEF)$

$X = 199.98912 * \cos(0°40'00")$

$X = \mathbf{199.97558}$

$Y = C * \sin(DEF)$

$Y = 199.98912 * \sin(0°40'00")$

$Y = \mathbf{2.32693}$

Exit Spiral (Ls2)

$a = (D * 100) / Ls2$

$A = (2°00'00" * 100) / 300.00$

$a = \mathbf{0.66667}$

$O = 0.0727 * a * ((Ls2 / 100)^3)$

$O = 0.0727 * 0.66667 * ((300.00 / 100)^3)$

$O = \mathbf{1.30861}$

$T = (Ls2 / 2) - (0.000127 * a^2 * (Ls2 / 100)^5)$

$T = (300.00 / 2) - (0.000127 * 0.66667^2 * (300.00 / 100)^5)$

$T = \mathbf{149.98628}$

$C = Ls2 - (0.00034 * a^2 * (Ls2 / 100)^5)$

$C = 300.00 - (0.00034 * 0.66667^2 * (300.00 / 100)^5)$

C = **299.96328**
DEF = (a * Ls2^2) / 60000
DEF = (0.66667 * 300.00^2) / 60000
DEF = **1.00000° or 1°00'00"**
Δs2 = 0.005 * D * Ls2
Δs2 = 0.005 * 2°00'00" * 300.00
Δs2 = **3.00000° or 3°00'00"**
U2 = C * Sin(Δs2 * 2 / 3) / Sin(Δs2)
U2 = 299.96328 * Sin(3°00'00" * 2 / 3) / Sin(3°00'00")
U2 = **200.02630**
V2 = C * Sin(Δs2 * 1 / 3) / Sin(Δs2)
V2 = 299.96328 * Sin(3°00'00" * 1 / 3) / Sin(3°00'00")
V2 = **100.02838**
X = C * Cos(DEF)
X = 299.96328 * Cos(1°00'00")
X = **299.91759**
Y = C * Sin(DEF)
Y = 299.96328 * Sin(1°00'00")
Y = **5.23508**

Main Curve

R = 5729.57795 / D
R = 5729.57795 / 2°00'00"
R = **2864.78898**

$\Delta m = \Delta t - \Delta s1 - \Delta s2$

$\Delta m = 36°29'16" - 2°00'00' - 3°00'00"$

$\Delta m = \mathbf{31°29'16"}$

$Lm = (\Delta m * R * \pi) / 180$

$Lm = (31°29'16" * 2864.78898 * \pi) / 180$

$Lm = \mathbf{1574.38889}$

$Tm = R * Tan(\Delta m / 2)$

$Tm = 2864.78898 * Tan(31°29'16" / 2)$

$Tm = \mathbf{807.62420}$

Note: See Book 4 - "Circular Curves" for formulas to solve other elements of the main curve.

Tangents

$Xa = Cos(\Delta s1) * (V1 + Tm)$

$Xa = Cos(2°00'00") * (66.67508 + 807.62420)$

$Xa = \mathbf{873.76668}$

$Ya = Sin(\Delta s1) * (V1 + Tm)$

$Ya = Sin(2°00'00") * (66.67508 + 807.62420)$

$Ya = \mathbf{30.51261}$

$Xb = Cos(\Delta m + \Delta s1) * (V2 + Tm)$

$Xb = Cos(31°29'16" + 2°00'00") * (100.02838 + 807.62420)$

$Xb = \mathbf{756.98547}$

$Yb = Sin(\Delta m + \Delta s1) * (V2 + Tm)$

$Yb = Sin(31°29'16" + 2°00'00") * (100.02838 + 807.62420)$

$Y_b = \mathbf{500.80556}$

$X_c = (Y_a + Y_b) / \tan(\Delta t)$

$X_c = (30.51261 + 500.80556) / \tan(36°29'16")$

$X_c = \mathbf{718.35573}$

$T_{s1} = X_a + X_b - X_c + U_1$

$T_{s1} = 873.76674 + 756.98547 - 718.35573 + 133.34112$

$T_{s1} = \mathbf{1045.73754}$

$T_{s2} = \sqrt{(X_c^2 + (Y_a + Y_b)^2)} + U_2$

$T_{s2} = \sqrt{(718.35575^2 + (30.51261 + 500.80556)^2)} + 200.02630$

$T_{s2} = \mathbf{1093.52165}$

Stationing

$SC(Sta) = TS(Sta) + L_{S1}$

$SC(Sta) = 2180+84.70 + 200.00$

$SC(Sta) = \mathbf{2182+84.70}$

$CS(Sta) = SC(Sta) + L_m$

$CS(Sta) = 2182+84.70 + 1574.39$

$CS(Sta) = \mathbf{2198+59.09}$

$ST(Sta) = CS(Sta) + L_{s2}$

$ST(Sta) = 2198+59.09 + 300.00$

$St(Sta) = \mathbf{2201+59.09}$

$PI(Sta) = TS(Sta) + T_{s1}$

$PI(Sta) = 2180+84.70 + 1045.74$

$PI(Sta) = \mathbf{2191+30.44}$

Note: The PI(Sta) is always calculated from the TS(Sta) and never from the ST(Sta).

Practical Example 2

At a minimum four (4) elements are required to calculate a full transitional spiral curve with unequal tangents. [1] The Total Delta Deflection (Δt), [2] Degree of Curvature (D) for the main curve, [3] Entrance Spiral length (Ls1) and [4] Exit Spiral length (Ls2). The Tangent to Spiral (TS) stationing is optional. This information can be derived from existing right of way plans, design plans and/or engineering design tables.

In order to calculate a full transitional spiral curve between two tangent lines, you first must determine the "Tangent In" and "Tangent Out" either by a survey of existing monumentation for existing centerlines or by a design alignment for a new centerline. It is beyond the scope of this book to describe the methods required to process existing monumentation to perform a centerline analysis for determination of the "Tangent In" and "Tangent Out".

For the purposes of this example, a Point on Tangent (POT) will be given on the "Tangent In", PI, a POT on the "Tangent Out", D, Ls1, Ls2 and TS(Sta) will be provided.

Given:

POT1 (a point on the Tangent In line)

N = 7128.34704

E = 7912.51319

PI

N = 5906.61886
E = 9197.04547
POT2 (a point on the Tangent Out line)
N = 4192.67915
E = 8482.47867
D = 3°00'00"
Ls1 = 250.00'
Ls2 = 350.00'
Solve for the following elements:
Δt = ??°??'??"
Ts1 = ????.?????
Ts2 = ????.?????
TS(Sta)
N = ??.??
E = ??.??
ST(Sta)
N = ??.??
E = ??.??

The solution can be found at the end of the book.

NOTES

TEN CHORD SPIRAL - RAILROAD SPIRAL CURVE

One of my first projects straight out of technical school was laying out spiral curves on a railroad project from Bill, Wyoming to Douglas, Wyoming using a Wild T2 theodolite and a 100' metal chain. Not having been formally trained on calculating spiral curves or calculus for that matter, I purchased a book titled "Railroad Curves and Earthwork" by C. Frank Allen, S.B. that gave me the necessary formulas to determine spiral curves for this railroad project. This book was expensive and hard to find back in those days but worth every penny. The cost of the book is just pennies in comparison.

Railroad Curves and Earthwork instructed me on the Ten Chord Spiral curve concept that the railroad uses and the process to properly layout spiral curves once the Tangent In and Tangent Out has been determined and staked in the field.

The spiral curves in "Railroad Curves and Earthwork" are based upon the Euler (Pronounced 'oil-er') Spiral Curve developed by Leonhard Euler. The Euler Spiral Curve is also the same mathematical model that is in use on most highway and railroad curves.

Keep in mind that there maybe transportation agencies that may use other mathematical models of spiral curves that are not covered in this book. I personally have not ran across any use of spiral curves other than the Euler Spiral mathematical model.

The subject matter in this book is entirely on the use of the Euler Spiral mathematical model.

Having said that, there is nothing magical or mythical about the Ten Chord Spiral Curve.

The only thing that the Ten Chord Spiral Curve does is divide the Entrance/Exit Spiral length into 10 equal segments that can be easily staked from the TS(Sta) or the ST(Sta) with a conventional instrument such as the Wild T2 and metal chain.

The other main difference between Highway and Railroad spiral curves is the Degree of Curvature. Highway curves uses an Arc Definition and the Railroad uses a Chord Definition. Book 4 - "Circular Curves" of this math series defines the differences between these two definitions.

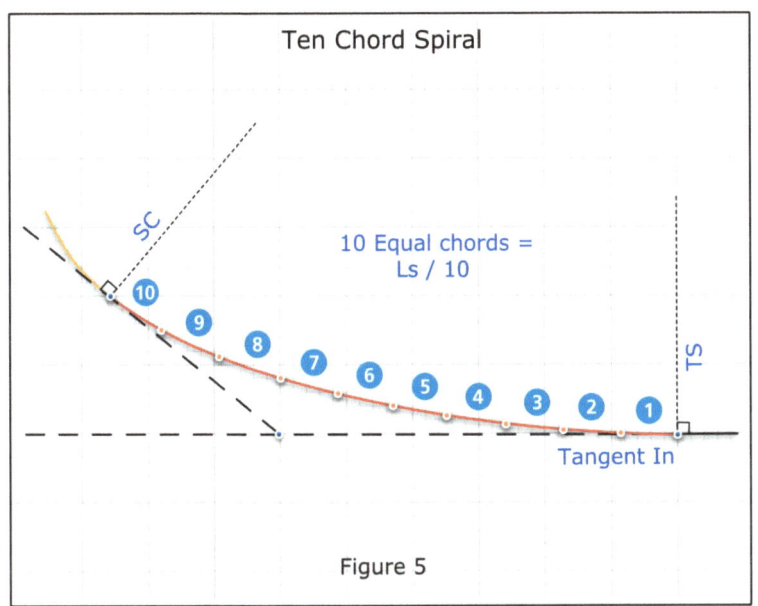

Figure 5

The spiral curve would be staked with the instrument setup at the TS station, back sighting along the tangent line then angles turned to each chord point along the curve. The chaining crew would pull a chain and measure the length of (LS / 10) to set each chord point along the curve then move ahead with each chord. This method will not be covered in this section of this book. There are plenty of publications available that cover that subject should you be interested in wanting to see how it is done.

With modern day survey equipment such as Total Station, Robotic Total Station and GPS, the staking of the centerline by setting up on the TS(Sta) and ST(Sta) and turning angles is seldom used.

POINTS ON A SPIRAL CURVE

What will be covered in this section is how to calculate a point along the spiral curve which can be 10 equal segments or any point at a particular station along the curve. The formulas are the same for either scenario. All you need to do is calculate the length along the spiral curve to each of the points that you want to solve for, apply this value to the formulas below to solve its position.

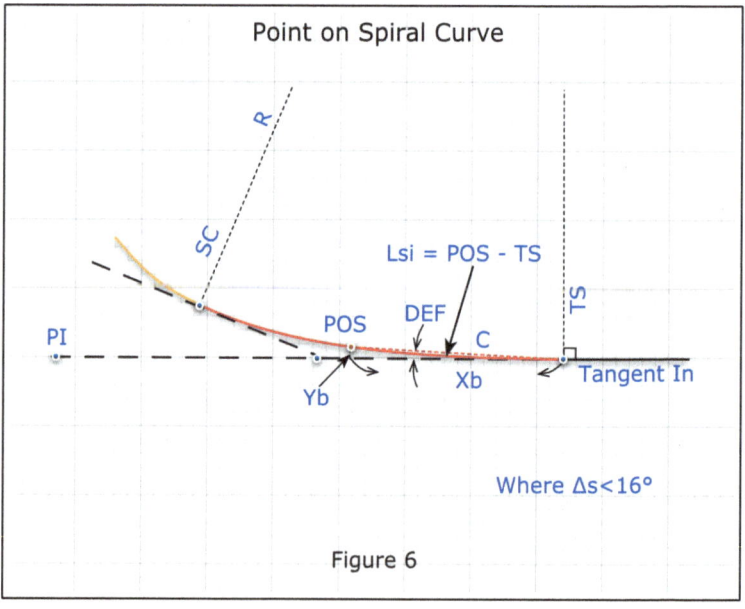

Figure 6

Quite often a point on a spiral curve will need to be determined. This can easily be calculated for any position along the spiral curve.

Refer to the definitions and formulas defined in the section "<u>Full Transition Spiral Curve - Equal Tangents</u>" that are in common with the following examples:

Additional Definitions:

POS = Point on Spiral Curve (Stationing)

Lsi = The distance along the spiral curve from the TS(Sta) to the POS

Xb = Distance along the X axis

YB = Distance along the Y axis

Additional Formulas:

Lsi = POS - TS(Sta)

C = Lsi - (0.00034 * a² * (Lsi / 100)⁵)

DEF = (a * Lsi²) / 60000

Xb = C * Cos(DEF)

Yb = C * Sin(DEF)

Example 3:

Given:

a = 1.00

TS(Sta) = 2180+84.70

POS = 2182+50.00

Solve for the following elements:

Point on Spiral Curve

Lsi = POS - TS(Sta)

$Lsi = 2182+50.00 - 2180+84.70$

$Lsi = \mathbf{165.30}$

$C = Lsi - (0.00034 * a^2 * (Lsi / 100)^5)$

$C = 165.30 - (0.00034 * 1.00^2 * (165.30 / 100)^5)$

$C = \mathbf{165.29580}$

$DEF = (a * Lsi^2) / 60000$

$DEF = (1.00 * 165.30^2) / 60000$

$DEF = \mathbf{0.45540° \text{ or } 0°27'19''}$

$Xb = C * Cos(DEF)$

$Xb = 165.29580 * Cos(0°27'19'')$

$Xb = \mathbf{165.29058}$

$Yb = C * Sin(DEF)$

$Yb = 165.29580 * Sin(0°27'19'')$

$Yb = \mathbf{1.31344}$

Practical Example 3

A new drainage culvert needs to be staked that is at a 30° skew to a spiral curve. Rectangular Coordinates are needed for the POS and end points of the culvert that can be staked with a GPS.

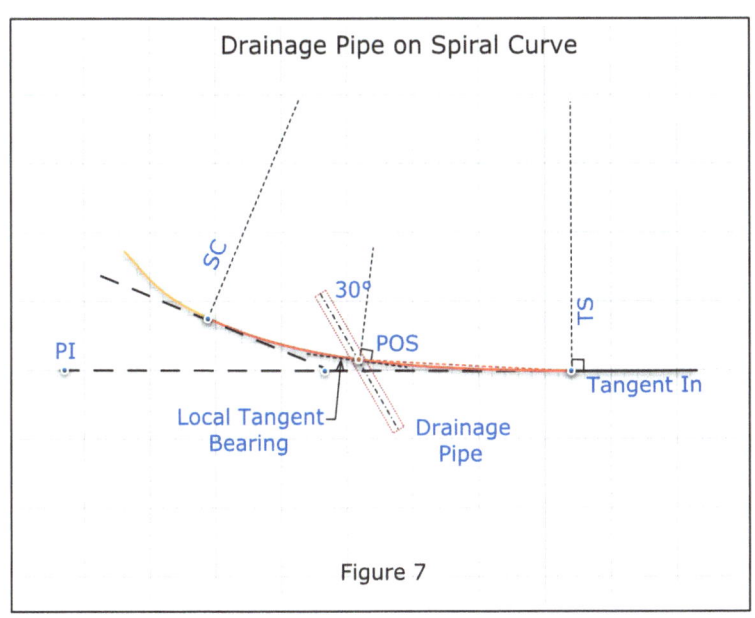

Figure 7

Given:

a = 0.50

Tangent In: N90°00'00"W

POS for Culvert= 253+12.45

Culvert Skew = 30° LT

Culvert Length = 50' RT and 50' LT

TS(Sta) = 10+00.00

TS(Sta) N = 10000.00000

TS(Sta) E = 20000.00000

Solve for the following elements:

POS N = ??.??

POS E = ??.??

End of Culvert RT side

N = ??.??

E = ??.??

End of Culvert LT side

N = ??.??

E = ??.??

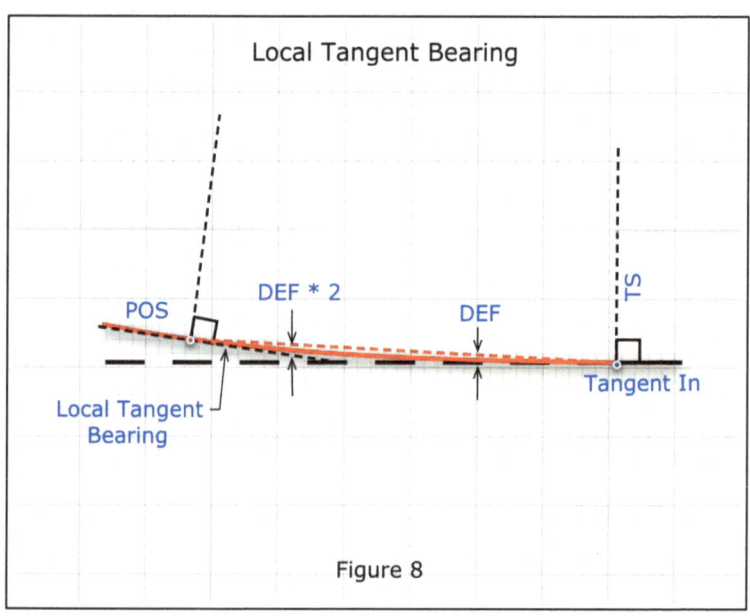

Figure 8

Hint: Determining the Local Tangent Bearing (LTB) at the POS will be required to solve this problem.

LTB = Tangent In Bearing +/- (DEF * 3) or

LTB = Chord Bearing +/- (DEF * 2)

The solution can be found at the end of the book.

NOTES

SOLUTIONS TO EXAMPLES

Note: Rounding error is dependent upon the number of decimal places that are utilized. It is recommended that at least 5 decimal places be used for all calculations then round the final answer as needed.

All angles must be converted to Decimal Degrees prior to performing trigonometric operations. See Book 1 - "Bearings and Azimuths" for methods on converting Degrees-Minutes-Seconds to Decimal Degrees and vice versa. Also see Book 1 for adding and subtracting bearings and angles.

Solution for Practical Example 1:

Given:

Tangent In: S70°12'24"W

Tangent Out: N64°17'36"W

D = 2°30'00"

Ls = 300.00'

TS(Sta) = 10+00.00

Solve for the following elements:

—SPIRAL CURVE—-

(Convert Tangent In and Tangent Out to Δt)

Δt = 180 - (Tangent Out + Tangent In)

Δt = 180 - (N64°17'36"W + S70°12'24"W)

Δt = **45°30'00"**

R = 5729.57795 / D

R = 5729.57795 / 2°30'00"

R = **2291.83118**

a = (D * 100) / Ls

a = (2°30'00" * 100) / 300.00

a = **0.83333**

Ts = (Tan(Δt / 2) * (R + O)) + T

    O = 0.0727 * a * ((Ls / 100)³)

    O = 0.0727 * 0.83333 * ((300.00 / 100)³)

    O = **1.63574**

    T = (Ls / 2) - (0.000127 * a² * (Ls / 100)⁵)

    T = (300.00 / 2) - (0.000127 * 0.83333² * (300.00 / 100)⁵)

    T = **149.97857**

Ts = (Tan(45°30'00" / 2) * (2291.83118 + 1.63574)) + 149.97857

Ts = **1111.70907**

C = Ls - (0.00034 * a² * (Ls / 100)⁵)

C = 300.00 - (0.00034 * 0.83333² * (300.00 / 100)⁵)

C = **299.94263**

DEF = (a * Ls²) / 60000

DEF = (0.83333 * 300.00²) / 60000

DEF = **1°15'00"**

Δs = 0.005 * D * Ls

Δs = 0.005 * 2°30'00" * 300.00

Δs = **3°45'00"**

—MAIN CURVE—

$\Delta m = \Delta t - \Delta s - \Delta s$

$\Delta m = 45°30'00" - 3°45'00" - 3°45'00"$

$\Delta m = \mathbf{38°00'00"}$

$Lm = (\Delta m * R * \pi) / 180$

$Lm = (38°00'00" * 2291.83118 * \pi) / 180$

$Lm = \mathbf{1520.00000}$

Note: See Book 4 - "Circular Curves" for formulas to solve other elements of the main curve as needed.

—STATIONING—

$SC(Sta) = TS(Sta) + Ls$

$SC(Sta) = 10+00.00 + 300.00$

$SC(Sta) = \mathbf{13+00.00}$

$CS(Sta) = SC(Sta) + Lm$

$CS(Sta) = 13+00.00 + 1520.00000$

$CS(Sta) = \mathbf{28+20.00}$

$ST(Sta) = CS(Sta) + Ls$

$ST(Sta) = 28+20.00 + 300.00$

$ST(Sta) = \mathbf{31+20.00}$

$PI(Sta) = TS(Sta) + Ts$

$PI(Sta) = 10+00.00 + 1111.70907$

$PI(Sta) = \mathbf{21+11.71}$

Solution for Practical Example 2:

Given:

POT1 (a point on the Tangent In line)

N = 7128.34704

E = 7912.51319

PI

N = 5906.61886

E = 9197.04547

POT2 (a point on the Tangent Out line)

N = 4192.67915

E = 8482.47867

D = 3°00'00"

Ls1 = 250.00'

Ls2 = 350.00'

Solve for the following elements:

Δt = ??°??'??"

You first need to solve for the bearings of "Tangent In" and "Tangent Out" lines then determine the deflection angle at the PI utilizing these bearings.

Lat = $N_2 - N_1$

Dep = $E_2 - E_1$

Bearing = Arctan(Dep / Lat) [Arctangent is also known as Tan^-1]

Note: See Book 3 - "Inverse Between Rectangular Coordinates" for formulas to solve for the Tangent In and Tangent Out bearings using the given coordinates above.

Lat1 = 5906.61886 - 7128.34704

Lat1 = **-1221.72818**

Dep1 = 9197.04547 - 7912.51319

Dep1 = **1284.53228**

Lat2 = 4192.67915 - 5906.61886

Lat2 = **-1713.93971**

Dep2 = 8482.47867 - 9197.04547

Dep2 = **-714.56680**

Tangent In = Arctan(1284.53228 / -1221.72818)

Tangent In = **S 46°26'08" E**

Tangent Out = Arctan(-714.56680 / -1713.93971)

Tangent Out = **S 22°37'55" W**

Δt = Tangent In + Tangent Out

Δt = S 46°26'08" E + S 22°37'55" W

Δt = **69°04'03"**

There are several steps required to solve the Northing and Easting for **Ts1** and **Ts2** as follows:

Entrance Spiral (Ls1)

a = (D * 100) / Ls1

a = (3°00'00" * 100 / 250.00

a = **1.20000**

C = Ls1 - (0.00034 * a^2 * (Ls1 / 100)^5)

C = 250.00 - (0.00034 * 1.20000^2 * (250.00 / 100)^5)

C = **249.95219**

$\Delta s_1 = 0.005 * D * L_{s1}$

$\Delta s_1 = 0.005 * 3°00'00" * 250.00$

$\Delta s_1 = \mathbf{3°45'00"}$

$U_1 = C * \mathrm{Sin}(\Delta s_1 * 2 / 3) / \mathrm{Sin}(\Delta s_1)$

$U_1 = 249.95219 * \mathrm{Sin}(3°45'00" * 2 / 3) / \mathrm{Sin}(3°45'00")$

$U_1 = \mathbf{166.70091}$

$V_1 = C * \mathrm{Sin}(\Delta s_1 * 1 / 3) / \mathrm{Sin}(\Delta s_1)$

$V_1 = 249.95219 * \mathrm{Sin}(3°45'00" * 1 / 3) / \mathrm{Sin}(3°45'00")$

$V_1 = \mathbf{83.37030}$

Exit Spiral (Ls2)

$a = (D * 100) / L_{s2}$

$a = (3°00'00" * 100) / 350.00$

$a = \mathbf{0.85714}$

$C = L_{s2} - (0.00034 * a^2 * (L_{s2} / 100)^5)$

$C = 350.00 - (0.00034 * 0.85714^2 * (350.00 / 100)^5)$

$C = \mathbf{349.86880}$

$\Delta s_2 = 0.005 * D * L_{s2}$

$\Delta s_2 = 0.005 * 3°00'00" * 350.00$

$\Delta s_2 = \mathbf{5°15'00"}$

$U_2 = C * \mathrm{Sin}(\Delta s_2 * 2 / 3) / \mathrm{Sin}(\Delta s_2)$

$U_2 = 349.86880 * \mathrm{Sin}(5°15'00" * 2 / 3) / \mathrm{Sin}(5°15'00")$

$U_2 = \mathbf{233.42734}$

$V_2 = C * \sin(\Delta s_2 * 1 / 3) / \sin(\Delta s_2)$

$V_2 = 349.86880 * \sin(5°15'00" * 1 / 3) / \sin(5°15'00")$

$V_2 = \mathbf{116.76813}$

Main Curve

$R = 5729.57795 / D$

$R = 5729.57795 / 3°00'00"$

$R = \mathbf{1909.85932}$

$\Delta m = \Delta t - \Delta s_1 - \Delta s_2$

$\Delta m = 69°04'03" - 3°45'00" - 5°15'00"$

$\Delta m = \mathbf{60°04'03"}$

$T m = R * \tan(\Delta m / 2)$

$T m = 1909.85932 * \tan(60°04'03" / 2)$

$T m = \mathbf{1104.15830}$

Tangents

$X a = \cos(\Delta s_1) * (V_1 + T m)$

$X a = \cos(3°45'00") * (83.37030 + 1104.15830)$

$X a = \mathbf{1184.98601}$

$Y a = \sin(\Delta s_1) * (V_1 + T m)$

$Y a = \sin(3°45'00") * (83.37030 + 1104.15830)$

$Y a = \mathbf{77.66809}$

$X b = \cos(\Delta m + \Delta s_1) * (V_2 + T m)$

$X b = \cos(60°04'03" + 3°45'00") * (116.76813 + 1104.15830)$

Xb = **538.71154**

Yb = Sin(Δm + Δs1) * (V2 + Tm)

Yb = Sin(60°04'03" + 3°45'00") * (116.76813 + 1104.15830)

Yb = **1095.65105**

Xc = (Ya + Yb) / Tan(Δt)

Xc = (77.66809 + 1095.65105) / Tan(69°04'03")

Xc = **448.80977**

Now that the various components of the spiral curve have been determined, the **Ts1** and **Ts2** can be solved.

Ts1 = ????.?????

Ts1 = Xa + Xb - Xc + U1

Ts1 = 1184.98601 + 538.71154 - 448.80977 + 166.70091

Ts1 = **1441.58869**

Ts2 = ????.?????

Ts2 = $\sqrt{}$(Xc² + (Ya + Yb)²) + U2

Ts2 = $\sqrt{}$(448.80977² + (77.66809 + 1095.65105)²) + 233.42734

Ts2 = **1489.65503**

The final step for this practical problem is to create rectangular coordinates for the TS(Sta) and ST(Sta). To accomplish this process you will utilize the PI Rectangular Coordinates, Tangent In & Tangent Out bearings, Ts1 & Ts2 distances.

Note: See Book 2 - "Create Rectangular Coordinates" for formulas to solve for the rectangular coordinates for the TS(Sta) and ST(Sta) points along the tangent lines.

TS(Sta)

The bearing from the PI along the back tangent will be 180° opposite of the "Tangent In" bearing that was determined at the beginning of this exercise.

Back Tangent = N 46°26'07" W

N = ??.??

N = PI northing + (Cos(Back Tangent) * Ts1)

N = 5906.61886 + (Cos(46°26'08") * 1441.58869)

N = **6900.11856**

E = ??.??

E = PI Easting - (Sin(Back Tangent) * Ts1)

E = 9197.04547 - (Sin(46°26'08") * 1441.58869)

E = **8152.47078**

Note: Use the proper algebraic signs for the Lat and Dep when using bearings. See Book 2 - "Create Rectangular Coordinates" for instructions on determining the correct algebraic signs and tips for using North Azimuth.

ST(Sta)

The bearing from the PI along the front tangent is the same as the "Tangent Out" bearing that was determined at the beginning of this exercise.

Front Tangent = S 22°37'55" W

N = ??.??

N = PI northing - (Cos(Front Tangent) * Ts2)

N = 5906.61886 - (Cos(22°37'55") * 1489.65503)

N = **4531.67350**

E = ??.??

E = PI Easting - (Sin(Front Tangent) * Ts2)

E = 9197.04547 - (Sin(22°37'55") * 1489.65503)

E = **8623.81134**

This practical example demonstrated the process to solve for the unequal tangent distances and rectangular coordinates for the beginning of the spiral curve TS(Sta) along the incoming tangent line and the end of the spiral curve ST(Sta) along the outgoing tangent line.

Additional components for the spiral curve such as the SC(Sta), RP & CS(Sta) can be computed and rectangular coordinates determined as needed.

Solution for Practical Example 3:

Given:

a = 0.50
Tangent In: N90°00'00"W
POS for Culvert = 12+53.45
Culvert Skew = 30° LT
Culvert Length = 50' RT and 50' LT
TS(Sta) = 10+00.00
TS(Sta) N = 10000.00000
TS(Sta) E = 20000.00000

Solve for the following elements:

POS N = ??.??
POS E = ??.??

Lsi = POS - TS(Sta)
Lsi = 12+53.45 - 10+00.00
Lsi = **253.45**
C = Lsi - (0.00034 * a² * (Lsi / 100)⁵)
C = 253.45 - (0.00034 * 0.50² * (253.45 / 100)⁵)
C = **253.44111**
DEF = (a * Lsi²) / 60000
DEF = (0.50 * 253.45²) / 60000
DEF = **0.53531 or 00°32'07"**
Xb (DEP) = C * Cos(DEF)

Xb = 253.44111 * Cos(00°32'07")

Xb = **253.43005**

Yb (LAT) = C * Sin(DEF)

Yb = 253.44111 * Sin(00°32'07")

Yb = **2.36770**

POS N = TS(Sta) N + Yb

POS N = 10000.00000 + 2.36770

POS N = **10002.36770**

POS E = TS(Sta) E - Xb

POS E = 20000.00000 - 253.43005

POS E = **19746.56995**

Determine the bearing of the centerline of the drainage pipe (DP).

Chord Bearing = Tangent In - DEF

Chord Bearing = N90°00'00"W - 00°32'07"

Chord Bearing = **N89°27'53"W**

LTB at POS = Chord Bearing - (DEF * 2)

LTB at POS = N89°27'53"W - (00°32'07" * 2)

LTB at POS = **N88°23'39"W**

Drainage Pipe Bearing = N88°23'39"W - 90° for perpendicular line to LTB + 30° LT for skew angle

Drainage Pipe (DP) Bearing = **N28°23'39"W**

End of Culvert RT side

N = ??.??

E = ??.??

LAT = Distance to the Right * Cos(DP Bearing)

LAT = 50.00 * Cos(28°23'39")

LAT = **43.98485**

N = POS N + LAT

N = 10002.36770 + 43.98485

N = **10046.35255**

DEP = Distance to the Right * Sin(DP Bearing)

DEP = 50.00 * Sin(28°23'39")

DEP = **23.77673**

E = POS E - DEP

E = 19746.56995 - 23.77673

E = **19722.79322**

End of Culvert LT side

N = ??.??

E = ??.??

LAT = Distance to the Left * Cos(DP Bearing)

LAT = 50.00 * Cos(28°23'39")

LAT = **43.98485**

N = POS N - LAT

N = 10002.36770 - 43.98485

N = **9958.38285**

DEP = Distance to the Left * Sin(DP Bearing)

DEP = 50.00 * Sin(28°23'39")

$DEP = \mathbf{23.77673}$

$E = POS\ E + DEP$

$E = 19746.56995 + 23.77673$

$E = \mathbf{19770.34668}$

NOTES

CONCLUSION

Spiral Curves can be complex mathematical formulas that only a calculus genius can comprehend but they don't have to be. This book has laid out a simple process to calculate most spiral curve combinations that you will likely come across in your surveying and engineering career using non-calculus methods to solve.

Computer programs have been designed and written that makes easy work for these type of calculations. In order to appreciate the mathematical beauty of spiral curves you are encouraged to manually calculate one every now and then.

See Book 7 - "The Myth about Spiral Curve Offsets" that will put the myth to rest about "there is no such a thing as a true spiral curve offset". After an endeavor starting back in 1976, I have finally derived a set of formulas that will calculate a true offset spiral curve to the centerline spiral curve. Despite what other survey/engineer professionals have written or said "there is no such a thing as a true spiral curve offset", deep inside me I knew that it could be done.

Here is a link to exhibits and a Spiral Curve Program.

http://www.cc4w.net/products.html

ABOUT THE AUTHOR
Jim Crume P.L.S., M.S., CFedS

My land surveying career began several decades ago while attending Albuquerque Technical Vocational Institute in New Mexico and has traversed many states such as Alaska, Arizona, Utah and Wyoming. I am a Professional Land Surveyor in Arizona, Utah and Wyoming. I am an appointed United States Mineral Surveyor and a Bureau of Land Management (BLM) Certified Federal Surveyor. I have many years of computer programming experience related to surveying.

This book is dedicated to the many individuals that have helped shape my career. Especially my wife Cindy. She has been my biggest supporter. She has been my instrument person, accountant, advisor and my best friend. Without her, I would not be the professional I am today. Cindy, thank you very much.

Other titles by this author:

http://www.cc4w.net/ebooks.html

www.ingramcontent.com/pod-product-compliance
Lightning Source LLC
Chambersburg PA
CBHW040849180526
45159CB00001B/360